Radioactive Waste-free Power Reactor

By Gordon L. Ziegler

Rad Waste-free Power Reactor

by Gordon L. Ziegler
Electrino Energy

Cover photo: Not much would be visible of a Rad Waste-free Power Reactor
installation except the property, because the accelerators would be underground.
If all goes well, this is the property where we wish to put the first Rad Waste-free
Power Reactor. (Photo taken by Gordon L. Ziegler January 1, 2009.)

This book was printed in the United States of America.

Rev. date: 01/22/2014

Author:
Gordon L. Ziegler dba
Electrino Energy
P.O. Box 1162
Olympia, WA 98507-1162 USA
ben_ent100@msn.com

To order additional copies of this book, contact:
Xlibris LLC
1-888-795-4274
www.Xlibris.com
Orders@Xlibris.com
541245

PREFACE

The world desperately needs a new inexpensive reliable safe energy source—an energy source that can eliminate our dependence on foreign oil. The energy source described in this little book is just such a new energy source. The author has found the secrets of unlocking the E = mc^2 energy in any matter. Reactor fuel need not be Uranium, Plutonium, or Hydrogen. It could be sand, water, air, garbage, toxic chemicals, or high level radioactive wastes—just to get rid of them. A little bit of fuel goes an extremely long way.

The type of reactor that can do all these things is the Electrino Fusion Power Reactor. Electrinos are a better fit in particle physics than quarks, and more symmetric. Electrinos have the interesting property that they can fuse to become new particles—but each time they do so, they switch from matter to antimatter, or vice versa. We take advantage of this phenomenon to annihilate any matter to obtain energy for electricity generation.

Some of the discoveries in this book are as old as 1983. Some came in 1995. Others came in 2004. The discovery, however, that made the EFP Reactor efficient enough to be self-sustaining and viable did not come until October 2007. The principle reactions, however, were not understood until July, 2008. Thus the EFP Reactor is a new discovery.

The EFP Reactor, however, is not a single source of energy. Two kinds of energy sources are required in EFP Reactors.

One type of energy produced is entropy energy for electricity production. The other type of energy produced with the reactor is positive order energy. This type of energy is necessary to make the accelerators 100% efficient and keep the irradiated electric panels from degrading with the strong radiation. This miracle working type of energy

3

also reverses aging, disease, and decay processes in people, animals, vegetation, minerals, and nucleons. A machine called Refresher 1 would produce the field that would do all these things by reversing the order to disorder arrow in the second law of thermodynamics in a controllable large area. The EFP reactor, if it is to be efficient enough to operate, must be placed within the active field of the Refresher 1. The theory of operation, specifications, costs, and impacts of Refresher 1 are described separately in the little book, *Refresher 1 Manual*. The specific theory of operation, description, specification, costs, and impacts of the EFP Reactor are described in this little book.

CONTENTS

Chapter 1

RAD WASTE-FREE REACTOR

1. The Need of a Radioactive Waste-free Reactor

Nuclear reactors do not give off carbon gases, and so could help fight global warming. But they have such nasty, dangerous radioactive wastes, which stay active and dangerous for centuries. The nuclear reactors themselves are dangerous, susceptible to terror, earthquake, and other hazards.

The need is great for a new kind of reactor that doesn't have radioactive wastes. In this book is theorized just such a reactor. If only there were no radioactive wastes, we could put up with lower power from the reactor. But pound for pound of fuel, the new reactor is actually 1000 times more powerful and fuel efficient. And if only there were no radioactive wastes, we could afford to have the new reactor cost more than the standard nuclear reactors. But in actuality, the new style reactor would cost sixty seven times less than a standard nuclear reactor for comparable or more electricity output.

What kind of reactor is it that has no radioactive wastes, has 1000 times the fuel efficiency, and less than one sixty seventh the costs, that doesn't need Uranium or Plutonium or Hydrogen for fuel? The next chapter gives the theory behind such a reactor.

Chapter 2

Theory of Operation

1. Introduction

The technical name for the Rad Waste-free Reactor is the Electrino Fusion Power Reactor (EFP Reactor). Electrino is the author's name for tiny electric particles that compose all light, matter, and gravitons in the author's new Grand Unification Theory (GUT). The main difference between the Standard Model and the new GUT is that fracton charges in the GUT come in $\pm$ e, $\pm$ e/2, $\pm$ e/4, and $\pm$ e/8; whereas fracton charges in the Standard Model come in $\pm$ 2e/3 and $\pm$ e/3. The change in fracton charges did not lead to untenable particle structures. The author induced the structures of every known particle according to the scheme in the GUT. They all worked out all right. And whereas it takes 61 elementary particles to build known light and matter in the Standard Model, it takes only one according to the GUT. The GUT has deeper levels of symmetry and lower orbits. This chapter develops the features of the radioactive waste-free EFP Reactor using the new GUT.

2. Elementary Particle Fusion

In the new GUT (which, by the way, is called Electrino Fusion Model of Elementary Particles), the particles are held together by symmetrical orbits, not glued together by gluons. The quarks, with $\pm$ 2e/3 and $\pm$ e/3 fracton charges, do not lend themselves to stable, symmetrical orbits, but the electrinos, with $\pm$ e, $\pm$ e/2, $\pm$ e/4, and $\pm$ e/8 fracton charges, do. In the model, photons are composed of heavy positive and negative whole charges orbiting about each other, and traveling together at the speed of light; electrons are made up of like light half charges orbiting about each other; and pions are made up of two orbiting pairs of like light fourth charges orbiting about each other. Notice the symmetry. Notice the orbits. Notice the space between the particles. Notice the individuality of the particles—bound only by the speed of light barrier and orbital mechanics.

It is important to notice the velocities of the particles and their behaviors at those velocities. All fractons (called electrinos in the model) travel either just slightly faster than the speed of light, or significantly faster than the speed of light. The point is, they all travel faster than the speed of light. For the light ones, this affects their radii—making them imaginary. This affects their force. Whereas slow like-charges repel, faster than c like-charges attract. This affects the potential energy of particles. This makes deep potential wells at the top of potential hills for the potential energy of charged particles. This affects the perceived mass-energy of the particles—positive instead of negative.

Faster than c like-charges attract. Negatively charged like half charges traveling just faster than c orbit around each other forming electrons. If the electrons never collide with any other electrons—at least not with sufficient energies—the half particle inertias in them cause the half charges to orbit always opposite each other—never approaching each other. But if electrons collide with each other with over 938 MeV each, four half charges come near to each other. The four half charges are not all held opposite each other. They all attract each other. What will happen? One half charge from one electron will be attracted to one half charge from the other electron. Nothing will stop the half charges. They will travel until they contact each other. What happens then? They are like charged. They form a new particle with twice the half charge—in other words a whole charge. We could say the half charges fuse to a whole charge.

When high energy electrons collide, not only do two half charges from opposite electrons fuse, the other two half charges on the opposite side fuse. We have four half charges from two electrons fusing to two whole charges. What then?

It is profitable at this juncture to assign fracton or electrino structures to simple particles. Pions are composed of four positive fourth charges in the manner already explained in the abstract. Electrons are made of two light weight negative half charges. Neutrons are constructed of a heavy positive whole particle orbited by an electron. If the constituents of pions were fused to the constituents of electrons, it would be to positive electrons—positrons—antimatter. If the sub-particles of negative electrons were fused to the heavy whole core particles of neutrons, it would be to negative neutrons—antimatter. If we started with the opposite charges of above, the particles would fuse to matter instead of antimatter. Every time there is a fusion of electrinos, there is a switch from matter to antimatter or visa versa.

What would happen to the negative half charges in electrons fused to whole particles above? The half charges would be negatively charged matter. The whole charges would be negatively charged core particles of antimatter—anti-protons and anti-neutrons. The anti-core-particles would scavenge from the graviton sea the remaining portions of anti-protons and anti-neutrons. The resultant anti-protons and anti-neutrons would drift into local protons and neutrons and annihilate them, giving off gamma rays, which could be converted into electricity. This is the foundation of the science of the radioactive waste-free EFP Reactor. The electricity comes from processed gamma rays, which come from the annihilation of protons and anti-protons and neutrons and anti-neutrons, which come from anti-protons and anti-neutrons, which come from negative heavy whole core particles (antimatter), which come from the fusion of half particles in electrons, which come from the collision electrons above

938 MeV each electron, with like spins in the center of mass frame.

3.   Efficiencies

Before electrons can have fusion of their half particles, they must be accelerated to at least the masses of protons—938.27231 ± 0.00028 MeV [1]—roughly at least 939 MeV. That is a necessary energy investment into the process. When the particles fuse, there follows an annihilation of both a proton and an anti-proton or a neutron and an anti-neutron.   Nearly twice as much energy in gamma rays results as was invested in the acceleration of electrons. At first this sounds good.   But then we realize we must be more than 50 per cent efficient over-all in order to be self-sustaining and be an energy source using this energy phenomenon.   That is hard to achieve.   State of the art accelerator efficiency in 1988 was itself only 50% [2]. While individual steam turbine efficiencies were as high as 96.1%, the world record steam turbine gross efficiency recently was 48.5% [3].   That is an overall efficiency for our process of less than 24.25%.   And we need 50% to break even, let alone have a surplus to become a new power source!

4.   A Surprising Turn

The lack of necessary efficiency of the fusion-annihilation reaction is discouraging. The author put this process on the back burner until he would receive greater light upon the subject.   Things took a surprising turn. Through fusing the sub-particles of positive electrons—positrons—in theory, he learned how to reverse the order to disorder arrow in the second law of thermodynamics. That is huge! That is a way to reverse aging, disease, and decay processes—to make old people young again and back out all diseases from existence! Let us read what he first wrote about the process and the phenomenon.

"The explanation that is usually given as to why we don't see broken cups gathering themselves together off the floor and jumping back onto the table is that it is forbidden by the second law of thermodynamics. This says that in any closed system disorder, or entropy, always increases with time. In other words, it is a form of Murphy's law: Things always tend to go wrong! An intact cup on the table is a state of high order, but a broken cup on the floor is a disordered state. One can go readily from the cup on the table in the past to the broken cup on the floor in the future, but not the other way round.

"The increase of disorder or entropy with time is one example of what is called an arrow of time, something that distinguishes the past from the future, giving a direction to time." [4]

5. Electrino Model and 2nd Law

The natural tendency of leptons in beta decay is that the parent lepton combines with one or more gravitons to produce more particles. In all natural reactions, the order energy of the resultant particles is less than or equal to the order energy of the original particles.

1.   Negative Energies.   Let us consider antimatter more carefully. "In the Dirac theory also, *the permissible energy values for a free particle range from* $+mc^2$ *to* $+4$ *and from* $-mc^2$ *to* $-4.$ The first of these results is of course just what we expect for a free particle—that its total energy

can have any value greater than its rest energy. But the second result is quite puzzling, since it implies the existence of states of *negative total energy*." [5]   Anderson in 1932 discovered positrons in cosmic radiation.   These were regarded as Dirac's negative energy particles. "The first two solutions of the Dirac equation . . . clearly describe a free electron of energy E and momentum **p**. The two negative energy electron solutions . . . are to be associated with the antiparticle, the positron." [6]

However, in the annihilation it is not $(+mc^2)$ + $(-mc^2) = 0$, but $2mc^2$ is the result of annihilation. [7] There is something strange going on with the minus signs in these equations. The calculations are inconsistent.

Maybe there are two kinds of energy considered. One we can call entropy energy E_S. In the annihilation reaction, $\#+mc^2\# + \#-mc^2\# = 2mc^2$. Entropy energy is the higher value. The other energy is order energy E_O. In order energy the same reaction is $(+mc^2) + (-mc^2) = 0$.

Let us consider entropy energy and order energy for particle decay schemes. There are a few decay schemes where no negative order energy (anti-matter) is introduced in the right hand side of the decay schemes. In those few instances, the final order energy is equal to the initial order energy (when kinetic energy is taken into account). But in most cases, a trace of negative order energy (anti-matter) is introduced into the right side of the decay schemes. There is nothing on the left hand sides of the decay schemes to correspond to this addition of a trace of negative order energy on the right sides of the decay schemes. Therefore, total order energy is less on the right hand sides of the decay schemes than on the left hand sides (if only by a trace). A few decay schemes introduce a lot of antimatter (as K⁻) on the right side of the decay scheme. The loss of order energy in the systems is greater in those cases. But in every case, for all natural processes, the order energy final is # the order energy initial, or

$$\Delta E_0 \leq 0. \qquad (1)$$

Let us check the order energy for electron electrino fusion reactions. Electrons made energetic by acceleration (as heavy as protons) fuse and form anti-protons. Matter is converted to anti-matter. Entropy energy is conserved, but not so order energy. Order energy is reduced in the extreme from +938 MeV to -938 MeV or more for each electron fused (two electrons are fused in each reaction). The order-disorder arrow for electron electrino fusion points in the usual direction. The system does obey the second law of thermodynamics.

2.   Reversing the Order to Disorder Arrow. What would happen if we fused the electrino constituents of positrons instead of the electrino constituents of electrons? Entropy energy E_S would again be conserved. Entropy would be increased. However, order energy E_O would go from -2 x 938 MeV to +2 x 938 MeV—from disorder to order. The order to disorder arrow would be reversed. This would be a reaction that would be prohibited by the second law of thermodynamics—unless the strong gravitational force that fuses the anti-semions would be stronger than the second law of thermodynamics

(which otherwise governs weak interactions). The stronger of the strong gravitational force and the second law of thermodynamics should be determined by experiment. More rides on that one experiment than perhaps on any one other experiment in this generation. If it is found that strong gravity is stronger than the second law of thermodynamics, then order can be restored at first in a small area, then for the whole earth.

Here we see that the entropy arrow of time and the order to disorder arrow of time are separate and distinct, and are not one and the same thing. While all the reactions the author has studied increase entropy, the fusion of positron anti-semions reverse the order to disorder arrow, making more order out of the disorder.

Positron constituent electrino fusion might not only take the electrinos from disorder to order. It could make other physical processes in a local area go from disorder to order. The positron fusion not only violates the second law of thermodynamics, it reverses the order to disorder arrow of that law in a local area, making other processes in that area reverse. Let us consider that process more to see how it might be regulated.

We guess the desired relationships for reversing the order to disorder arrow in the second law of thermodynamics through dimensional analysis. We want to solve for r, the maximum radius in which the reversed law would be effective. There is a way we can obtain a length from combinations of our variables and constants. That way is in the right hand side of Eq. (2). The whole expression is the thermodynamic relation we are seeking. The thermodynamic relation is:

$$(\Delta E_o)_t > 0 \; where \; r < \frac{(\Delta E_o)_1 \; c}{ik}, \qquad (2)$$

where E_o is the order energy–the positive or negative energy in the pair production of particles; ΔE_o is the change in the order energy, where $(\Delta E_o)_t$ is the change in the total order energy of the system, and where $(\Delta E_o)_1$ is the change in the order energy for a single source reaction—for a positron fusion reaction it is approximately 2×10^9 eV/collision x 1.6×10^{-19} joules/eV = 3.2×10^{-10} joules/collision; c is the speed of light—approximately 3.0×10^8 m/s; we shall solve for the effected radius r; i is the beam current in each beam in Coulombs per second (we will solve for 10^{-11}); k is the ratio of particle energy to particle charge. This energy per charge is the accelerated energy of the particle (roughly 1×10^9 ev times 1.6×10^{-19} joules/ev = 1.6×10^{-10} joules) divided by the charge of each positron (q = 1.6×10^{-19} coulombs), which equals 10^9 joules per coulomb. The collision efficiency eff is not needed in this equation, because the result is not in particles, but is already in collisions.

Incredibly, the lower the current, the bigger the radius of the affected area. And the greater the current, the smaller the radius of the effected area. With 10^{-11} A beam currents, the effected radius r solves for 9.6 meters—roughly 10 meters, which describes a small area—less than a tenth of an acre.

To get an idea of the positron beam currents needed to reverse the order to disorder arrow of the second law of thermodynamics in what size of affected radius, see Table 1 below.

For an area the size of   r   beam current

House	10 m	10 pA
four football fields	100 m	1 pA
community	1 km	100 fA
city	10 km	10 fA
Israel	160 km	0.6 fA
U.S.	2,400 km	0.04 fA
World	13,000 km	0.008 fA
Sun	1.7E11 m	6E-22 A

Table 1. Beam currents versus affected radius for reversal of the order to disorder arrow of the second law of thermodynamics.

Remarkably enough, the affected area of second law reversal calculates to increase with the reduction of positron beam current. Area control is merely a matter of timed gating of the positrons in the positron-positron collider. [8]

6.   Rate of Reversed Aging

The author will now calculate the rate at which reverse aging will occur in the calculable radius of the active Refresher: The beginning energy of the host particles (positrons) from which the fusion process takes place is $2m_ec^2$ per individual reaction. The ending energy of the host particles (protons) to which the fusion process tends is $2m_pc^2$ per individual reaction. $\dfrac{\Delta E_p}{\Delta E_{e^+}} = \dfrac{+2m_pc^2}{-2m_ec^2} \approx -1836$. This is a unit less expression from the available energy terms. What we seek is another unit less expression $\dfrac{\Delta t_r}{\Delta t}$, where t is the normal time during which a person or object ages, and t_r is the reverse time (negative) during which a person or object un-ages. The quotient is the relative rate of un-aging compared to aging. This also is a unit less quotient. What use of particle fusion parameters can yield such a unit less quotient? What terms are available to derive such a unit less quotient? What about the first terms and unit less quotient? If we equate them, we have $\dfrac{\Delta t_r}{\Delta t} \approx -1836$. Reverse time would be negative and 1836 times as fast as forward aging time. Forward aging of 100 years would be un-aged in 19.89 days. Forward aging of 1 year would be un-aged in approximately 4.77 hours of machine time.

7.   Miracle Working Power of the Refresher 1

The theoretical discovery of the order to disorder arrow in the second law of thermodynamics reverser (Refresher 1 for short) was a surprising turn, and engrossed the author for several years. By simply reversing the natural arrows between ordered events, many miraculous results were found to take place in theory.

What does it mean that the order to disorder arrow in the second law of thermodynamics is reversed? Events naturally come in order indicated by the arrows:

Healthy   young adult→aging→wrinkles→aging→cancer→death→ cremation→scattering ashes

Reversing the order to disorder arrow in the second law of thermodynamics means all the arrows between the ordered events are turned around. The old and diseased become young and healthy. The clock is not really reversed. Adults do not become children again and disappear to extinction. The system just tends to maximum order, which is at young adulthood. Children still grow up to maximum order at young adulthood.

Many similar reversals can occur in the animal kingdom and the environment. The author imagined many marvelous things, but virtually forgot about the EFP Reactor.

8.   EFP Reactor in the Field of the Refresher 1

Finally the thought came, "What would occur if the EFP Reactor were in the field of a Refresher? The concepts of the effects assembled slowly. The accelerator electronics would not have resistive heating in the field. As a result the accelerator would be room temperature superconductive. There would not be any need for cryogenic energy losses. The accelerator would be 100% efficient.

Reversing the order to disorder arrow in the second law of thermodynamics greatly affects all things with which we are familiar. But what would it do photovoltaic cells in a high energy gamma field? Outside the Refresher field, photovoltaic cells in the high energy gamma field would become damaged. They would become more and more damaged with time. This is a form of aging. What would happen if the aged photovoltaic cells were put in an order reversed Refresher field? The cells would un-age back to the original condition. What would happen if photovoltaic cells in an order reversing Refresher field were exposed to high level gamma radiation? They would not become damaged or aged. What would happen to the power that would ordinarily be absorbed in the aging process? Would it not be added to the power converted from radiation to electricity in the photovoltaic cells?

But what about the miscellaneous heating that would occur to photovoltaic cells in a high level radiation field outside an order reversing field of a Refresher? The heating process, though not necessarily damaging and aging, also occurs as an ordered process in the second law of thermodynamics. If the order to disorder reversed field of the Refresher were added, the photovoltaic cells would be cooled down. Heating would not occur in the field. What would happen to the power ordinarily lost to heating? Would not it be added to the power converted from radiation to electricity in the photovoltaic cells?

But what about the gamma photons that would not age the photovoltaic cells or heat them, but would pass through them without affecting them? What if the Refresher field were added, what would then take place? The next question can resolve this question. Is the shielding loss included in the order to disorder arrows in the reaction equations? Yes. Then with the addition of the Refresher field, the elusive photons would return or never penetrate the photovoltaic cells. What would happen to that power? Would not it be added to the power converted from radiation to electricity in the photovoltaic cells? This result is the hardest to take. We need experiment to settle this. If this paragraph were not true, we would expect it would take layers upon layers—many feet of photovoltaic cells piled on top of each other to stop the gamma photons. But if this paragraph is true, then gamma rays as well as sunlight could be stopped by a single layer of photovoltaic cells in the order to disorder in the second law of thermodynamics reverser of the Refresher. In the reversed field, the photovoltaic cells should be 100% efficient.

An EFP Reactor must be built and operate in the field of a Refresher.

While an individual photovoltaic cell may be 100% efficient, it would not be possible to cover every spot around the reactor with photovoltaic cells. But it should be possible to achieve 60% to on the order of 100% efficiency—enough for the source to be self-sustaining and an energy source.

9.   What about Radioactive Wastes?

As we now experience the second law of thermodynamics, neutrons + products $\rightarrow$ neutron activation products. Reverse that and activation products become deactivated and neutrons are given off. Another reaction involving neutrons: $n \rightarrow p + e + anti\ \nu_e$. Reverse that and neutrons are produced. In the field of the Refresher 1, neutrons appear stable. Also in the field, radioisotopes are all backed out of existence. As long as the Refresher 1 field is on, the EFP Reactor will be radioactive waste free.

References

[1] SUMMARY TABLES OF PARTICLE PROPERTIES, January 1, 1998, Particle Data Group, as quoted by *CRC Handbook of Chemistry and Physics, 80th Edition* (Boca Raton: CRC Press, 1999), pp. 11-1 to 11-49.

[2] SDI: technology, survivability, and software (Diane Publishing Co., May, 1988), p. 140, NTIS order #PB88-236245.

[3] Mathias Deckers, Steam Turbine Blading Technology for Siemens, Germany, "CFX AIDS DESIGN OF WORLD'S MOST EFFICIENT STEAM TURBINE," http://www.ansys.com/assets/testimonials/siemens.pdf.

[4] Stephen Hawking, *A Brief History of Time*--From the Big Bang to Black Holes (New York: Bantam Books, 1988), pp. 144, 145.

[5] Robert B. Leighton, *Principles of Modern Physics* (New York: McGraw-Hill Book Company, Inc, 1959), p. 665.

[6] Francis Halzen, Alan D. Martin, *Quarks and Leptons* (New York: John Wiley & Sons, 1984), p. 107.

[7] David S. Saxon, *Elementary Quantum Mechanics* (San Francisco: Holden-Day, 1968), p. 386.

[8] Gordon L. Ziegler, *Electrino Physics* (Lacey, Washington: Electrino Energy, 2010), Chapter 16, http://www.benevolententerprises.org/.

Chapter 3

HISTORICAL BACKGROUND

March 1964	As high school physics student, decide there is an aether after all (Einstein notwithstanding). Aspire to derive relativity in an aether.
1970	Write historical summary of physics discoveries.
1977	Derive special and general relativity in an aether.
1981	Found Benevolent Enterprises.
1981	Begin theoretical work on science of chonomics (particle structure).
1982	Write "Manipulating Gravity and Inertia Through a New Model of the Universe."
1983	Discover in theory electrino fusion power.
1983	See value of reversing second law of thermodynamics.
1984	See vision of world benefits of reversing the second law of thermodynamics.
1987	Begin paper on structure of elementary particles.

1990	Begin synthesizing scientific discoveries in manuscript *The Unified Universe*.
February 27, 1992	Received a letter from E. R. Siciliano, Ph.D. pointing out that existing incoherent particle beams can collide only about one in 10,000 particles accelerated. This letter had large impact on quest for electrino fusion power, apparently showing the need for coherent beams.
1994	Enlarge and redraft physics work under the title, *Formulating the Universe*, Volume 1, which in part contained work on the unified field theory and unified particle theory.
November 1994	Begin in earnest to theorize and design coherent electron and coherent positron sources to help in electrino fusion processes.
February 16, 1995	Founded Coherent Electron Source, LLC.
1995	Drafted *Formulating the Universe*, Volume 2. Learned how to reverse order to disorder arrow in the second law of thermodynamics through the fusion of positron anti-semions. Initially had incorrect r^2 model of second law reversal.

April 1, 1996   Conceived the basics of cavity
                          Electron source.

April-October, 1996 Worked together with JEOL USA
                          Inc., an electron microscope
                          company, to design, fabricate, and
                          test a substrate cavity electron
                          source.

1996-1999   Attempted to achieve coherent
                          electron source through various
                          methods of cavity fabrication and
                          testing.

2000   Learn of a U.S. patent for "A
                          Method and Apparatus for
                          Generating High Energy Coherent
                          Electron Beam and Gamma-Ray
                          Laser," by Hidetsugu Ikegami,
                          Takarazuka, Japan (U.S. Patent No.
                          5,887,008, Mar. 23, 1999).

2001   Begin draft of *Formulating the
                          Universe*, Volume 3.

August 2003   Disclose method of using positive
                          anti-semion fusion as a force field
                          bomb shelter. Was then off by a
                          factor of 10^4 in efficiency calcula-
                          tions.

1996-2004   Revise and correct *Formulating the
                          Universe*, Volumes 1 and 2, several
                          times.

December 2004   Discover that electrino fusion may
                          Be 6.24×10^{18} times as efficient as

e⁻ e⁺ collisions, because the magnetic and weak forces make the particles smart guided bombs to fuse with each other. If this is true, electrino fusion power generation may be possible as a new and cheaper source of electricity. If not, EFP generation may not be possible.

December 15, 2004	Completed version of FTU, Vols. 1 and 2, incorporating this question in Volume 2, Chapter 6. Saw super nova as pre-existing test of the high efficiency version of EFP.
December 2004	Prepared prospectus for securing Venture capital for EF2LTF (Electrino Fusion Second Law Test Facility) and EFMPPI (Electrino Fusion Model Power Plant Inventors).
August 2005	Maked contact with the one man that is qualified to and build the EF2LTF accelerator-collider, James M. Potter, Ph.D. of JP Accelerator Works, Inc., Of Los Alamos, New Mexico. Obtained from him an estimate of How much this would cost.
January 2006	Published "A New Way to Calculate Electron and Muon g/2-factors" in *Galilean Electrodynamics Journal*.
September 18, 2006	Submitted grant application to DOE for EF2LTF facility.

March 11, 2007	Finish *Electrino Physics* book, synthesizing, correcting, and extending materials in *Formulating the Universe*, Volumes 1-3.
October 2007	Discover field of Refresher 1 would greatly increase the efficiency of the EFP Reactor, making it possible.
October 16, 2007	Revised *Governing With Refresher 1*
November 2, 2007	Compiled information in *Refresher 1*.
June 4, 2008	Finished and submitted "Prediction of the Masses of Every Particle, Part 1."
July 2008	Learned principal reactions of EFP Reactor in field of Refresher 1. Wrote about radioactive wastes (or lack of them) in field of Refresher 1.

Chapter 4

ANTICIPATED RESULTS/
POTENTIAL COMMERCIAL APPLICATIONS

The Refresher 1, or EF2LTF (Electrino Fusion Second Law Test Facility), can operate without the EFP Reactor, but the EFP Reactor cannot be self sustaining without the Refresher 1 field. Whether the EFP Reactor and the Refresher 1 are first built, or whether the Refresher 1 is built alone, depends on how much money is available and how much is willing to be risked at first. But whatever is built first, these same three results are liable to occur with the construction and testing of the electrino fusion facilities.

1.   The electrino fusion facilities would test a Grand Unification Theory (GUT) for the benefit of science and mankind.

2.   The electrino fusion facilities would test the feasibility of the generation of power from the fusion of electron semions. This form of power generation could soon make up 50% or more of world-wide power generation—a huge commercial market. There may be a large market for low power generators—1 to 100 MW. There may also be a market for high power generators—1,000 to 10,000 MW. All these are possible by varying the beam currents and varying the depth of solar cells in the energy recovery areas.

3.   The electrino fusion facilities would also test the feasibility of reversing the order to disorder arrow in the second law of thermodynamics, thereby reversing aging, disease, and decay processes. The Refresher 1 is not only designed to make tests in a small local area, but treat larger and larger areas, until the entire earth is treated simultaneously. The book by the author, *Governing With Refresher 1*, foretells a wide range of effects from the

Refresher 1, affecting many disciplines. Chapter 16 in *Electrino Physics*, by the author, calculates the range of the field depending on the positron beam current strength. This type of electrino fusion would have more far reaching results than the electron semion EFP fusion results.

Since the Electrino Fusion Power Reactor is radioactive waste free when operated in the field of the Refresher 1, and since it is 1000 times more fuel efficient than nuclear power plants, not needing cooling towers, and since the cost of the Radioactive Waste-free Power Reactor is less than one sixty seventh that of nuclear power plants, the EFP reactors would undoubtedly quickly come in high demand. They do not depend on the wind or sunshine. They would be the ideal power source.

Chapter 5

DESCRIPTION AND SPECIFICATIONS

Phase 1

The Rad Waste-free Reactor is efficient, self sustaining, and rad waste free only in the field of the Refresher 1 (order to disorder arrow in the second law of thermodynamics reverser). See *Refresher 1 Manual* for description and specifications of the Refresher 1.

The Refresher 1 positron accelerator-collider should be housed in a separate containment building adjacent to the Rad Waste-free Reactor electron accelerator-collider containment building. The design of both containment buildings must be in harmony with the regulations concerning the design of containment buildings.

Phase 2

The electron accelerator-collider for the reactor should have nominal energy 1940 MeV variable ± 100 MeV. The current should be variable from 0 to 1.06 Amp (for 1.0 GW power maximum). The current should be pulsed at lower average currents and CW at maximum power.

In order to achieve electrino fusion, the electrons of one leg of the accelerator must be spin flipped before collision.

The fuel chamber has to be annular about the legs of the accelerator and the collider. The anti-protons will be created in the collider. By adjusting the energy of the accelerators, the anti-protons must be adjusted in energy so that they will penetrate the casing of the collider before annihilating protons and neutrons in the fuel chamber. Else the collider casing would be the fuel and would decompose, necessitating frequent rebuilding of the accelerators.

The electron accelerator collider must be surrounded by an inward focused shell of photo-voltaic cells to convert gamma rays to DC voltage. Without the Refresher 1, the photo-voltaic cells would not be very efficient. But with the Refresher 1, the photo cells should have about 1.0 efficiency. Without the Refresher 1, the photo cell leads should have limited current ratings. But with the Refresher 1, the photo cell leads should be super-conductive, accommodating the increased power with Refresher 1 operation.

The power source will require also a connective grid connecting all the photocells; a low voltage, high current port to the outside of the containment building; a 60-cycle modulator in phase with the local power grid; and a high voltage transformer to the public power grid. It should be noted here that super-conductivity should extend to the limit of the Refresher 1 active field (including the modulator and transformer and port). Therefore a 1.0 GW transformer could safely be performed by a 1.0 kW transformer in the field, etc.

Chapter 6

COSTS

Phase 1: Refresher 1

Accelerators	$33,000,000
Land, facility, personnel, misc. for 3.5 yr.	17,000,000

Phase 2: Power Source

Accelerators	38,000,000
Power Source equipment	1,000,000
Power Source facility	1,000,000
Land: included with Refresher 1	0
Personnel: included with Refresher 1	0
	$90,000,000

Chapter 7

TIME CONSTRAINTS

Micro-design of the accelerators	12 months
Pre-fabrication at remote site	18 months
Assembly and testing on site	6 months
Total time from funding to commissioning	36 months

Chapter 8

Inventor, Builder, Owner

Inventor

Gordon L. Ziegler, dba
Electrino Energy
PO Box 1162
Olympia, WA 98507-1162 USA
ben_ent100@msn.com
electrino.energy@yahoo.com

Builder

James M. Potter, Ph.D., Pres.
JP Accelerator Works, Inc.
2245 47th Street
Los Alamos, NM 87544 USA
505-690-8701
jpotter@jpaw.com
http://www.jpaw.com

Owner

Benevolent Enterprises
PO Box 1162
Olympia, WA 98507-1162 USA
ben_ent100@msn.com

www.ingramcontent.com/pod-product-compliance
Lightning Source LLC
Chambersburg PA
CBHW021049180526
45163CB00005B/2351